SERAMA CHICKEN FARMING FOR BEGINNER

A Step-By-Step Guide To Raising, Breeding, And Caring For The World's Smallest Hen With Tips On Feed, And Health Management

Holden bodhi

Contents

CHAPTER ONE 9
  Serama Beginner's Guide To Chicken Farming...............9
    Overview Of Serama Chickens 9
    A Synopsis Of The Serama Breed.......................10
    Distinctive Features And Attributes..................11
    The Advantages Of Keeping Serama Chickens 13

CHAPTER TWO..17
  Establishing A Serama Chicken Farm.....................17
    Selecting The Proper Site 17
    Crucial Tools And Materials..........................19
    Temperature Control And Insulation 23

CHAPTER THREE 25
  Selecting Serama Chickens..............................25
    Choosing Healthy Adults Or Chicks....................25
    Comprehending Breed Varieties And Standards..........27
    Where To Purchase Chickens From Serama 29

CHAPTER FOUR...33
  Providing Food For Your Serama Chickens 33
    The Dietary Needs For Development And Egg Production
....33
    Feed And Supplement Types............................36
    Portion Sizes And Feeding Schedules..................38

CHAPTER FIVE...41
  Serama Chickens' Well-Being And Health.................41

- Typical Serama Chicken Health Problems 42
- Vaccinations And Preventative Care 44
- Identifying Illness Symptoms 46

CHAPTER SIX 49
- Serama Chicken Breeding 49
  - Knowing Genetics And Breeding Practices 49
  - Formulating A Breeding Strategy 52
  - Fundamentals Of Incubation And Hatching 54

CHAPTER SEVEN 57
- Serama Chicken Housing And Environment 57
  - Keeping The Coop Clean And Safe 57
  - Considerations For Ventilation And Bedding 59

CHAPTER EIGHT 63
- Taking Care Of Chickens In All Seasons 63
  - Adapting Care To Seasonal Variations 63
  - The Coop's Winterisation 67
  - Controlling Summer Heat 70

CHAPTER NINE 75
- How To Care For And Train Serama Chickens 75
  - Fundamental Handling Methods 75
  - Getting Along With Others And Managing Your Chickens 77
  - Getting Ready For Competitions Or Shows 78

CHAPTER TEN 81
- Getting Ready And Competing 81

Getting Serama Chickens Ready For Performances.........81

Comprehending Show Criteria And Evaluation 84

Advice On Winning At Contests..86

CHAPTER ELEVEN..89

The Promotion And Sale Of Serama Chickens....................89

Investigating New Markets For Your Chickens................89

Formulating Successful Advertising Plans......................92

Regulations And Legal Aspects 95

Copyright © 2024 by Holden bodhi

All rights reserved.

No part of this publication may be reproduced, distributed, or transmitted in any form or by any means, including photocopying, recording, or other electronic or mechanical methods, without the prior written permission of the publisher, except in the case of brief quotations embodied in critical reviews and certain other non commercial uses permitted by copyright law.

DISCLAIMER

The information provided in this book, is intended for educational and informational purposes only. The content is based on research, personal experiences, and general knowledge about farming. It is not intended to substitute professional advice or expert consultation. Readers are encouraged to seek professional guidance when implementing any practices or techniques discussed in this book.

The author and publisher make no representations or warranties of any kind regarding the accuracy, applicability, or completeness of the contents of this book. Any reliance you place on such information is strictly at your own risk. The author and publisher shall not be held liable for any damages, losses, or injuries resulting from the use of the information provided.

Additionally, the author does not endorse, recommend, or affiliate with any individual, product, service, website, organization, or brand mentioned or referenced in this book. Any such references are solely for informational purposes, and no warranty or guarantee is implied. The inclusion of these references does not imply any endorsement or partnership by the author.

By reading this book, you acknowledge and accept that the author and publisher are not responsible for any consequences arising from your use of the information provided.

CHAPTER ONE

Serama Beginner's Guide To Chicken Farming

Overview Of Serama Chickens

Keeping hens, especially the adorable Serama breed, may be a fulfilling endeavor. Serama chickens are one of the tiniest kinds of chickens and are also highly favored by poultry aficionados.

They are native to Malaysia. They are a desirable option for both novice and seasoned farmers due to their small size, distinct personalities, and stunning plumage. We will delve into the intriguing realm of Serama chickens in this tutorial, covering their background, traits, and advantages of keeping them.

A Synopsis Of The Serama Breed

Often called the "world's smallest chicken breed," Serama chickens weigh between 13 and 19 ounces as adults. They are ideal for backyard or urban gardening because of their compact size, which enables them to flourish in smaller areas. Serama chickens give visual appeal to any flock because they come in a variety of colors and feather patterns. Their distinctive appearance—broad breasts, short back, and upright posture—accompanies their small size. Because of their remarkable look, they are frequently featured for their grace and beauty at exhibitions and poultry shows.

The breed's inception dates back to Malaysia in the late 1800s when it was developed for both competition and companionship. Originating from the Malay word "serama," which means "to strut," the moniker "Serama" is an appropriate term for these pompous little birds. The breed has developed over time, giving rise to several variants, including both feathered and non-feathered ones. With devoted clubs and organizations promoting their care and breeding, the Serama breed has achieved international renown.

The amiable and gregarious nature of Serama hens is one of its distinctive features. Seramas are renowned for being gentle and easy to handle, which sets them apart from many other chicken breeds and makes them a great option for families with young children. Because they may grow extremely connected to their owners and frequently love human interaction, their friendly nature helps explain why they are so popular as pets.

Distinctive Features And Attributes

Serama chickens differ from other breeds due to several distinctive traits. Their size, which varies greatly from person to person, is one of their most distinctive characteristics. Breeders can choose birds that fit their preferences thanks to the weight classifications of Serama hens, which range from A class (up to 13 ounces) to D class (up to 19 ounces).

Apart from their size, Serama chickens have several kinds of feathers, such as silkied, frizzled, and smooth feathers. Because each bird can display its distinctive feathering, this diversity contributes to the breed's appeal. Any farmer wishing to establish a flock that is visually pleasing can choose from a variety of feather colors, such as black, blue, red, white, and more.

The versatility of Serama chickens is yet another outstanding trait. They are surprisingly resilient birds that can survive in a variety of climes, despite their diminutive size. Although they can withstand both heat and cold, their well-being must provide them with sufficient cover and protection from harsh weather.

Because of their versatility, they can be used in a variety of farming settings, including both rural and urban backyards.

Moreover, in contrast to larger varieties, Serama chickens require less upkeep. They are a cost-effective option for novices because of their small size, which means they need less space and feed. They are also easier to manage during regular care tasks like feeding, watering, and health examinations due to their peaceful nature. This trait is especially desirable for new farmers who could be put off by bigger, more aggressive breeds.

The Advantages Of Keeping Serama Chickens

There are several advantages to raising Serama hens, which makes them a desirable option for novices. First and foremost, they are perfect for urban or suburban environments where space is limited because of their tiny size, which enables them to be stored in constrained areas. People who do not have the land to raise larger poultry breeds now have more opportunities because of this accessibility.

Serama chickens are renowned for their ability to lay eggs in addition to their space-saving benefits. Serama hens can lay small, edible, and ornamental eggs, although they are not as prolific layers as larger breeds. Usually weighing around one ounce, their eggs can be a delightful treat for family and friends or a fun addition to the kitchen.

Additionally, Serama hens make wonderful pets. They are fun companions because of their playful attitude and amiable disposition. They frequently form close relationships with their owners and thrive on social engagement. Because Serama hens can teach kids about responsibility and animal care in a hands-on way, this feature is especially appealing to families with kids.

The ability to compete in poultry shows and competitions is a major advantage of growing Serama chickens. Serama owners can display their birds at different events with the right breeding and care, which can be a rewarding experience for lovers. Among chicken farmers, these contests frequently promote a feeling of camaraderie by offering chances for networking, experience sharing, and mutual learning.

Finally, it is impossible to ignore the beauty of Serama hens. Any farm or backyard environment is made more beautiful by its striking appearance, range of colors, and distinctive feather types. Those who want to improve their outside area might use a flock of Serama chickens as a decorative and practical addition.

All things considered, Serama chickens are a fun breed with many benefits for novice producers. Serama chickens are a great option for anyone wishing to start a poultry farming business because of their distinctive traits, outgoing personality, ability to compete, and adaptability to compact settings. Serama chickens will add happiness and satisfaction to your farming experience, regardless of whether you choose to raise them as pets, for egg production, or display.

CHAPTER TWO

Establishing A Serama Chicken Farm

There are several important factors to take into account while starting a Serama chicken farm to provide these endearing birds with a comfortable environment. This guide will take you through all the necessary processes, from picking the ideal spot to creating a cozy coop.

Selecting The Proper Site

To guarantee the health and welfare of your birds, it is essential to choose the ideal site for your Serama chicken farm. The perfect location should offer lots of room, adequate drainage, and weather protection.

Accessibility

Take accessibility into account when picking a place. For everyday care duties like feeding, watering, and egg collection, your chicken farm should be conveniently located. This will motivate you to routinely check on the health of your birds and spend more time with them. An excessively remote location may result in neglect and possible health problems.

Shade and Sunlight

Serama hens require a balance of shade and sunlight and do best in mild climates. To keep the coop warm and promote natural behaviors, the area should ideally get some direct sunlight throughout the day. To prevent your birds from overheating in the sweltering summer months, it should also have parts that are shaded. Awnings and tree planting are two ways to get the shade you need.

Defence Against Predators

Serama hens are at serious risk from predators. It's crucial to pick a spot that is naturally shielded from common predators like foxes, raccoons, and raptors. Think about choosing a spot with natural obstacles like hedges or fencing. To better protect your birds, you should create a safe coop and run.

Drainage and Soil

Consider the sort of soil in the area you have selected. To keep water from collecting around the coop, which can cause bacterial development and harm your hens' health, you need well-draining soil. Consider raising your coop or picking a different location if the area has poor drainage or is prone to flooding. Maintaining a clean and dry environment is essential to keeping your flock disease-free.

Crucial Tools And Materials

The next step after choosing the ideal site is to collect the necessary tools and materials for rearing Serama hens. Purchasing high-quality products will increase your farm's productivity and the general well-being of your birds.

Coop and Run Chickens

The most important part of your chicken farm is the coop. It should give your Serama chickens a safe spot to roost and lay their eggs as well as protection from inclement weather. The coop should be large enough to comfortably house your flock. In the coop, each bird should have at least 2 to 4 square feet of room, and in the outside run, 8 to 10 square feet.

Features like roosting bars for sleeping, nesting boxes for laying eggs, and sufficient ventilation to guarantee fresh air circulation should all be taken into account when planning your coop. To keep your hens safe from predators while still enabling them to walk and feed, the outside run needs to be properly fenced.

Supplies for Watering and Feeding

Your Serama hens' well-being depends on proper diet and hydration. To guarantee that your birds always have access to food and clean water, spend money on high-quality feeders and waterers. Waterers and feeders that operate automatically can reduce waste and simplify everyday chores. Make sure the feeders are the right size for your chickens; the petite Serama breed does best with smaller feeders.

Material for Bedding

Maintaining comfort and hygiene in the coop requires careful selection of the bedding material. Common choices for insulation and moisture absorption include straw, hay, or wood shavings. Changing the bedding regularly will help keep odors under control and lower the chance of illness. Furthermore, a dry and clean atmosphere will improve the general health of your hens.

Medical Supplies

It's important to keep your flock healthy. Stock up on first aid kits, prescription drugs, and other necessary health supplies. Keep an eye out for any symptoms of sickness in your hens and seek advice from a veterinarian if necessary. Maintaining your birds' health not only increases their lifespan but also keeps their egg-laying cycle productive.

Creating a Cosy Coop Design

For your Serama chickens to live in a secure and cozy environment, a well-designed coop is necessary. To guarantee the contentment and efficiency of your flock, the design should prioritize comfort, security, and usefulness.

Layout and Space

Give layout and space top priority while designing your coop. Make certain that each bird has adequate space to go around without being crowded. Think about designating distinct spaces in the coop for feeding, roosting, and nesting. This isolation will encourage natural activities like dust bathing and foraging while also lowering stress levels.

Airflow

To keep the coop environment healthy, proper ventilation is essential. A healthy airflow will assist in lowering humidity and stop dangerous gases like ammonia from building up. To control temperature and airflow, install windows or vents that may be opened or closed as necessary. Nonetheless, throughout the winter months, make sure vents are positioned to avoid drafts.

Temperature Control And Insulation

Due to their small size, Serama hens may be susceptible to severe weather conditions. The coop will remain stable with insulation, staying cool in the summer and warm in the winter. Think about utilizing insulating materials like foam board or bales of straw. You can also supply fans or heat lamps to regulate the temperature as needed.

Accessibility of Cleaning

Consider cleaning accessibility when designing the coop. Make sure you have easy access to every part of the coop for regular maintenance, and use materials that are simple to clean and maintain. By including features like detachable dropping trays, you can make cleaning easier and give your hens a clean living environment.

Features of Enrichment

Lastly, to keep your Serama chickens cognitively engaged, think about incorporating enrichment elements into the coop and run. Dust baths, climbing frames, and perches will promote natural behaviors and increase your birds' general well-being. A healthy and productive flock can be fostered by creating a stimulating atmosphere.

CHAPTER THREE

Selecting Serama Chickens

Selecting the appropriate birds is one of the first and most crucial choices you will make when beginning a chicken farming business in Serama. Serama chickens are a popular choice for home farmers and chicken enthusiasts due to their small size, amiable nature, and endearing appearance. To ensure a fruitful and fulfilling farming experience, it is essential to choose the healthiest and highest-quality Serama chicks. Choosing healthy chicks or adults, comprehending breed standards and variety, and knowing where to get Serama chickens are all covered in this part.

Choosing Healthy Adults Or Chicks

Choosing whether to start with chicks or adults is the first step in selecting Serama chickens. Depending on your farming objectives and degree of experience, each choice provides a unique combination of benefits and difficulties.

• Selecting Chicks: Choosing chicks can be a wonderful way to get to know your birds and see them develop if you're new to chicken farming or Serama chickens. Consider a chick's general health and energy while choosing her. Serama chicks in good health should be aware of their environment, active, and alert. Look for clean, fluffy down and eyes that are clear and bright. Steer clear of girls who seem drowsy, have drooping wings, or exhibit respiratory distress symptoms such as sneezing or wheezing. Check the cleanliness of their vents because clogged or dirty vents may be a sign of digestive problems. Additionally, healthy girls should not have any physical abnormalities or leg issues.

• Choosing Adults: Choosing healthy adult Serama chickens is just as crucial for anyone wishing to start their Serama farming with older birds. Examine the general health of the mature chickens when assessing them. Despite their diminutive size, seramas should not be extremely thin or emaciated and instead have a complete, well-rounded body. Examine their feathers; birds in good health will have smooth, glossy feathers free of broken or missing plumage. Additionally, look for scaly leg mites or any indications of swelling on their legs. As a good measure of general health, the bird's comb should be firm and brilliant red, free of shriveling or discoloration.

Comprehending Breed Varieties And Standards

Although Serama chickens are available in a large range of colors and feather kinds, it's crucial to understand breed standards, particularly if you intend to display your birds or breed them for sale. The weight, posture, and appearance of Serama chickens are among the standards recognized by the American Poultry

Association (APA) and the American Bantam Association (ABA).

• Size and Weight: Serama hens are the tiniest breed of chicken in the world, and their weight should be commensurate with that. Females should weigh slightly less, about 10 to 14 ounces, while adult males should ideally weigh between 12 and 16 ounces. In competitive show settings, birds larger than these would not be regarded as authentic Seramas.

• stance: The proud, erect stance of Serama chickens is one of their most distinguishing traits. The stance of a well-bred Serama should be nearly vertical, with a high chest and a steeply angled tail. To give them a majestic look, their wings should droop low and touch the ground.

• Feathers: Serama hens have a variety of feather types, such as silkied, frizzled, and smooth-feathered. While frizzled Seramas have feathers that curl outward from their bodies, giving them a distinctive and fluffy appearance, smooth-feathered Seramas have conventional, sleek feathers. Like the Silkie chicken breed, Silkied Seramas have feathers that are silky and fur-like. All kinds of feathers can be excellent pets, but depending on your preferences and the objectives of your Serama farming, some types can be more suited for breeding or exhibition.

Whether you're choosing birds for competitive exhibiting or personal enjoyment, knowing the breed standards will help you choose birds that are both healthy and fit breed requirements.

Where To Purchase Chickens From Serama

Finding a reliable source to get your Serama chickens is the next step after you have a better understanding of the breed's requirements and know what to look for in healthy birds. Serama chickens can be purchased in a variety of ways, each with unique benefits.

• Local Breeders: Purchasing straight from a local breeder is among the finest ways to guarantee that you're getting healthy Serama chickens. You will be able to visit the flock, ask questions, and select your birds in person if the breeder is trustworthy. Before bringing the birds home, you can use this hands-on method to confirm their quality and health. Enquire about the birds' ancestry and any breeding procedures the breeder uses to keep healthy, standard-compliant hens when you buy from them.

• Online Hatcheries: Online hatcheries can be a practical choice if there are no local breeders in your area. Numerous hatcheries may deliver chicks right to your home and specialize in unusual and exotic varieties like Seramas. Even while buying chicks online can be easy, it's crucial to study the hatchery to make sure you're getting from a reliable one. Before making a purchase, look for reviews, testimonials, and health assurances. To give you greater choice over how your flock looks, some hatcheries even let you choose particular feather colors or kinds.

• Poultry exhibitions and Expos: Meeting breeders and purchasing Serama chicks can also be accomplished by going to poultry exhibitions and expos. You can meet knowledgeable breeders, get a close-up look at a range of Seramas, and ask questions about raising and caring for this unusual species at these events. You may purchase premium Seramas from renowned breeders at many exhibitions that also sell birds.

• Social Media Groups and Forums: Online communities devoted to Serama chickens have been more well-known in recent years. Finding birds for sale can be made easier with the help of websites, forums, and social media groups devoted to poultry aficionados. These platforms are used by numerous breeders and enthusiasts to sell chicks and adult birds, frequently sending them to clients all over the nation. When purchasing from online groups, exercise caution, though, as it may be more difficult to confirm the birds' health and quality without physically viewing them. Always request pictures, health guarantees, and prior customer reviews.

You can lay the groundwork for a fruitful and pleasurable Serama chicken farming experience by choosing healthy Serama chickens from reliable suppliers and being aware of breed requirements. Making sure your birds are healthy, and content, and achieve your farming objectives is crucial, regardless of whether you select adults or chicks, or if they have smooth or frizzled feathers.

CHAPTER FOUR

Providing Food For Your Serama Chickens

One essential component of your Serama hens' care that has a direct impact on their development, well-being, and egg production is feeding them. Maintaining an active, nutritious, and productive diet is crucial for your hens. The dietary needs of Serama hens, the kinds of feed and supplements that are available, and the ideal feeding times and serving amounts to maximize their health will all be covered in this section.

The Dietary Needs For Development And Egg Production

For optimum growth and egg production, serama chickens, like all poultry, have particular nutritional requirements that must be satisfied. Proteins, carbs, lipids, vitamins, and minerals are the main ingredients of their diet.

The proteins

For Serama hens to grow and thrive, proteins are necessary. They are essential for the development and repair of bodily tissues, and they are especially vital for hens that lay eggs during the chick stage. While adult layers should be fed a diet with 16–18% protein, growing chicks should be fed a diet with 20–24% protein. Commercial layer feeds, soybean meal, and insects are good sources of protein.

Fats and Carbohydrates

For hens, the main energy source is carbohydrates. They should make up a sizable amount of the diet and be mostly obtained from grains like wheat and maize. Although needed in modest amounts, fats give off energy and facilitate the absorption of fat-soluble vitamins. To keep a healthy weight, make sure fats are consumed in moderation.

Minerals and Vitamins

For Serama hens to be healthy and productive overall, vitamins and minerals are essential. A, D3, E, and the B-complex vitamins are important vitamins. Because they aid in the production of eggshells, minerals like calcium, phosphorus, and sodium are also crucial, particularly for chickens that produce eggs. Generally speaking, giving your hens a full-layer feed guarantees that they get the vitamins and minerals they need.

Water

Finally, there must always be access to clean, fresh water. Digestion, vitamin absorption, and general health all depend on water. Make sure your hens always have access to water, especially in warmer weather when they require more hydration.

Feed And Supplement Types

To satisfy your Serama chickens' nutritional needs, you must select the proper kind of feed. There are numerous meals and supplements available, each with a specific purpose and life stage in mind.

Initial Feed

A beginning feed is necessary for chicks. This feed is specifically designed to promote quick growth and development and has a high protein content (around 20–24%).

To support robust immune systems, starter meals frequently include extra vitamins and minerals.

Feed for Growers

Your Serama hens will switch to grower feed as they become older. This meal is designed for chickens aged six weeks to twenty weeks and normally has about 18% protein. It helps children grow healthily and gets them ready for adulthood.

Feeding Layers

Your chickens need a layered meal with 16–18% protein and extra calcium once they begin laying eggs. The purpose of layer feeds is to promote egg production and guarantee robust eggshell formation.

Add-ons

Although premium commercial feeds often offer a well-rounded diet, several supplements can improve your Serama hens' nutritional intake. These may consist of:

• Calcium Supplements: Crushed oyster shells are one type of calcium supplement that can be given separately to laying hens to guarantee proper calcium intake.

- Vitamin Supplements: Vitamin supplements can strengthen the immune system and promote general health under stressful or illness-related situations.

- Probiotics: Probiotics can enhance digestion and support gut health. They might be especially helpful if your hens have intestinal problems.

Portion Sizes And Feeding Schedules

The health of your Serama chickens depends on setting up a feeding schedule and figuring out portion proportions. By controlling feeding schedules and amounts, you can keep your hens healthy and productive by preventing overeating and obesity.

Feeding Timetable

Your hens' eating habits can be controlled with a regular feeding regimen. Chicks require frequent feedings to promote their rapid growth, so give them food and water constantly.

When your hens get older, think about feeding them according to this schedule:

• Morning: In the morning, provide a freshly prepared portion of feed. Since they are most active during the day, chickens will profit from eating in the morning.

• Afternoon: If you feed them twice a day, give them a second serving of food. This can guarantee that they get enough nourishment.

• Evening: Make sure your hens have access to enough feed to last the entire evening if you choose to free-feed them.

Sizes of Portion

The age and quantity of hens in your flock will determine the size of your portions. A broad rule of thumb is:

• Chicks: Give each chick one ounce of feed daily at first, then progressively more as they become bigger.

- Growers (6 weeks to 20 weeks): Give each bird four to five ounces of feed every day.

- Laying Hens: Depending on body condition and egg output, give each hen 5–6 ounces of layer feed daily.

You can keep your Serama hens healthy and productive by keeping an eye on their bodily condition and modifying portion quantities as needed. To avoid spoiling and to keep the feeding area hygienic, make sure that any food that is left over is taken out after a few hours.

In conclusion, the general health and productivity of your Serama hens depend on feeding them a balanced diet that is suited to their growth and egg production requirements. You can provide your flock with the finest care possible by being aware of their nutritional needs, selecting the appropriate feed and supplements, and setting up a regular feeding plan.

CHAPTER FIVE

Serama Chickens' Well-Being And Health

Keeping Serama hens healthy and happy is essential to their longevity and production. Knowing the different facets of chicken health as a novice farmer will help you raise a successful flock. Common health problems, preventative treatment, and identifying symptoms of disease in Serama hens are covered in this section.

Typical Serama Chicken Health Problems

Like any other breed of chicken, Seramas, which are renowned for their small size and distinctive appearance, can be susceptible to specific health problems. Comprehending these possible issues is crucial for prompt action and guaranteeing the general welfare of your flock.

1. Respiratory Conditions: Infectious bronchitis and avian influenza are among the most prevalent respiratory conditions affecting Serama hens. Breathing difficulties, nasal discharge, coughing, and sneezing are possible symptoms. Since these illnesses spread swiftly, it's critical to separate afflicted birds right once and get veterinary advice for diagnosis and treatment.

2. Coccidiosis: This intestinal tract parasite infection is especially prevalent in young chicks. Lethargy, bloody diarrhea, and decreased appetite are some of the symptoms. Keeping housing clean and using medicated feed are two preventative methods that can greatly lower the risk of coccidiosis.

3. Worm Infestation: Worms can lead to several health problems for Serama hens, such as anemia, poor feather quality, and weight loss. Frequent deworming is crucial, and you should

speak with a veterinarian to determine the best deworming regimen and medicine.

4. Mites and Lice: Serama hens may become infested with external parasites such as mites and lice, which can cause discomfort, feather loss, and anemia. Infestations can be avoided by routinely inspecting your birds for indications of these pests and keeping their living space clean. When necessary, eradicate these pests with appropriate treatments such as poultry dust or diatomaceous earth.

5. Leg Issues: Serama chickens are susceptible to leg issues, such as sprains and fractures, because of their diminutive size. Creating a roomy and secure space will reduce the chance of harm. To avoid joint tension, make sure the bedding in their housing is soft.

Vaccinations And Preventative Care

The secret to keeping your Serama hens healthy is preventative maintenance. A proactive approach to health management can greatly lower disease risk and enhance general well-being.

1. Frequent Health Checks: To spot any possible problems early, perform frequent health checks on your Serama chicks. Look for behavioral abnormalities, unusual droppings, or indications of respiratory discomfort. Effective therapy depends on early discovery.

2. Vaccinations: Vaccinations are essential for keeping chickens healthy. Marek's disease, Newcastle disease, and infectious bronchitis vaccinations are frequently administered to Serama hens. To create a vaccination regimen that is suitable for your flock, considering their age and health, speak with your veterinarian.

3. Strict biosecurity measures should be put in place to guard against disease in your flock. Avoid contact with wild birds, keep your chicken coop clean, and restrict entrance. Make sure

the new hens you add to your flock are kept in quarantine for a minimum of two weeks to keep an eye out for any symptoms of the disease.

4. Nutrition: For Serama hens to be healthy, their nutrition must be balanced. Make sure they have access to fresh water and premium feed that is tailored to meet their requirements. To support optimum health, add fruits, vegetables, and proteins to their diet.

5. Stress Reduction: Your Serama chickens' health may suffer as a result of stress. Provide enough room, shelter, and enrichment activities in a safe and comfortable living setting. Their immune system and general health will be preserved by lowering stressors.

Identifying Illness Symptoms

Being able to spot symptoms of disease in your flock is crucial for quick action as a Serama chicken farmer. Your birds' health results might be greatly impacted by early diagnosis and treatment.

1. Behavioural Changes: Keep an eye out for any behavioral changes, such as decreased activity, lethargy, or separation from the flock. Since chickens are gregarious creatures, their withdrawal could be a sign of a more serious health problem.

2. Eating and Drinking Patterns: Keep an eye on your hens' eating and drinking patterns. A sharp decline in thirst or appetite may indicate a medical condition. A bird should be examined by a veterinarian if it stops eating or drinking for longer than a day.

3. Physical Symptoms: Keep a look out for any outward manifestations of disease, including unusual droppings, alterations in the condition of the feathers, swelling, or discharge from the eyes and nose. Any of these symptoms need to be treated right away.

4. Keep an eye out for respiratory symptoms, such as wheezing, sneezing, or coughing. Isolate the afflicted bird and get veterinary advice for additional assessment if you see any of these symptoms.

5. Frequent Record-Keeping: You can spot trends and possible issues by keeping thorough records of your flock's health, including immunization history, health concerns, and treatment outcomes. For future use and when speaking with a veterinarian, this knowledge is priceless.

In conclusion, taking care of your Serama hens' health and well-being entails being aware of prevalent health problems, practicing preventative care, and spotting symptoms of the disease. You may guarantee a prosperous and healthy environment for your Serama hens by continuing to be proactive and aware of the demands of your flock.

CHAPTER SIX

Serama Chicken Breeding

Knowing Genetics And Breeding Practices

The intriguing task of growing Serama chickens necessitates a thorough comprehension of the genetics and breeding techniques. Serama chickens are well-liked in the poultry industry because of their petite stature, amiable nature, and eye-catching look. But to properly breed them, one needs to understand the fundamentals of heredity and genetics.

Genetic Characteristics

Serama chickens' morphological and behavioral characteristics are mostly determined by their genetic makeup. Genetic factors influence traits like temperament, body size, and feather color. Anyone wishing to breed Serama chickens must comprehend dominant and recessive features. In contrast to a recessive trait, which only manifests when both parents possess it, a chick who inherits a dominant characteristic from one parent will exhibit that trait.

Comparing Line And Cross Breeding

There are two methods for breeding Serama chickens: cross-breeding and line-breeding. To preserve a pure line and improve desired qualities, line breeding entails marrying closely related birds. Although there are hazards associated with this approach, including the possibility of genetic problems brought on by inbreeding, it can be useful in maintaining certain traits that are exclusive to the Serama breed.

Conversely, cross-breeding is the process of marrying unrelated birds to increase genetic variety and introduce new features. This strategy may result in stronger, healthier birds by removing genetic problems brought on by inbreeding. On the other hand, it may also produce children that fall short of the breed's expectations.

Choosing Breeding Stock

For Serama chicken breeding to be successful, selecting the appropriate breeding stock is essential. Take into account the birds' temperament, health, and conformity to breed standards while choosing them for your breeding program. Seek out birds that exhibit desirable qualities including pleasant dispositions, bright feather colors, and good conformation.

Maintaining thorough records of the performance and provenance of your breeding stock is also crucial. You can use these records to track the effectiveness of your breeding program over time and make well-informed judgments about potential breeding pairings.

Formulating A Breeding Strategy

For Serama chicken breeding to be successful, a well-organized breeding plan must be created. A breeding plan describes your objectives, the desired traits, and the precise breeding techniques you'll employ to achieve them.

Setting Objectives

Establish your breeding objectives first. Are you trying to increase certain physical characteristics, increase egg output, or produce show-quality birds? Having specific goals can help you make judgments during the breeding process. For instance, pay attention to characteristics like feather color, body form, and general health if your objective is to raise show-quality birds.

Selecting Pairs for Breeding

It's time to select your breeding pairings after you've decided on your objectives. Determine which birds best fit your requirements by evaluating your current stock. Make sure the selected pairs complement each other genetically and take into account variables like age, health, and temperament.

Seasonality and Timing

Another crucial element of any breeding strategy is timing. Serama hens usually breed seasonally, with spring and summer being the busiest times for mating. To improve the likelihood of successful fertilization and hatching, schedule your breeding operations around this natural cycle.

Tracking Development

Regularly evaluate your progress towards your objectives during the breeding process. Maintain thorough records of matings, hatch rates, and offspring characteristics. This information will help you stay on course to meet your goals and make necessary changes to your breeding plan.

Fundamentals Of Incubation And Hatching

For Serama chickens, successful incubation and hatching are essential stages in the reproduction process. Your chicks' health and survival can be greatly impacted by how well you handle these stages.

Comprehending Incubation

During incubation, the ideal conditions for growing embryos are maintained. The optimal incubation temperature for Serama chickens is approximately 99.5°F (37.5°C) with a humidity level of 50–60%. To precisely monitor these parameters, use a dependable incubator that has a temperature and a hygrometer.

Placing the Eggs

Make sure the pointed end of the eggs is facing down when you place them in the incubator. The developing embryo can move more easily in this position. Turning the eggs frequently—roughly three times a day—during incubation is also essential to keep the embryo from adhering to the shell.

Egg Candling

You can check for embryo development and fertility by candleing the eggs around the seventh day of incubation. A strong light is shone through the egg during this procedure to look for any indications of life, like blood vessels or movement. To keep the incubation environment healthy, throw away any eggs that are not viable or sterile.

Time to Hatch

Serama chicks usually hatch 21 days following the start of incubation. Keep a careful eye on the humidity levels as the hatch date draws near; they should rise to roughly 65-70% to aid in the hatching process. Allow the chicks to fully emerge from the incubator after they start to hatch, as this process may take several hours.

After-Hatching Treatment

After hatching, give the chicks a cozy, secure place to live. Throughout the first few weeks of life, make sure they have access to food and water and keep a careful eye on their health. Your Serama chicks will develop into healthy, energetic birds that are prepared to be used as pets or added to your breeding program with the right care and attention.

CHAPTER SEVEN

Serama Chicken Housing And Environment

Raising Serama hens in a healthy and happy environment requires proper housing and surroundings. These tiny birds, who are renowned for their amiable disposition and beautiful looks, need a clean, safe, and cozy environment to flourish. The numerous housing and environmental factors that are essential for Serama hens will be discussed in this section.

Keeping The Coop Clean And Safe

How Important a Clean Coop Is

For Serama hens to be healthy and happy, their coop must be kept clean. Frequent cleaning keeps pests, parasites, and dangerous pathogens from growing and endangering your birds. Start by creating a coop with nesting boxes and detachable perches that are simple to clean.

Weekly and Daily Cleaning Schedules

A hygienic atmosphere can be maintained by establishing a daily regimen for spot-cleaning any droppings. Every week, give the entire coop a thorough cleaning. To get rid of any residues, remove the bedding, scrub the surfaces with a light disinfectant, and then rinse well.

Methods of Pest Control

Use pest management techniques to keep insects and rodents out of your coop. Employ traps or natural deterrents, like diatomaceous earth, which works well against pests but is harmless for hens. Additionally, to prevent drawing undesirable animals, make sure that feed is kept in sealed containers.

Frequent Examinations

Develop the practice of routinely checking the coop for wear or damage. Make sure windows and doors are sealed tightly and keep an eye out for openings that pests could exploit. To keep the environment safe, take quick care of any problems.

Considerations For Ventilation And Bedding

Selecting the Proper Mattress

To keep a coop atmosphere healthy, bedding material is essential. Straw, wood shavings, or shredded paper are good choices for Serama chickens. Choose according to your preferences and availability, as each style has advantages and disadvantages. For example, wood shavings are less likely to draw insects than straw, which is both absorbent and insulating, but may harbor pests.

Management of Bedding

To facilitate absorption and insulation, bedding should be kept at least 4 inches deep. To preserve aeration and lessen odors, fluff the bedding regularly. Soiled bedding should ideally be changed once a week, but in warmer months when decomposition is accelerated, it may need to be changed more regularly.

Needs for Ventilation

A healthy coop habitat depends on having enough ventilation. Proper ventilation lowers the chance of respiratory problems in hens and avoids the accumulation of moisture. Make sure the bottom parts of your coop are draft-free and have vents close to the top to let warm air out.

Seasonal Factors

Make sure your coop is adequately insulated throughout the winter months while still permitting adequate airflow. Heat lights should be used sparingly to prevent overheating. To assist control your hens' temperature and avoid heat stress throughout the summer, create shady spots outside the coop.

Enrichment and Outdoor Space

Establishing an Outdoor Area

Serama hens love being outside because it gives them the chance to explore and behave naturally. Give them access to a safe outdoor run or free-range space where they may graze, scratch, and enjoy the sunshine. To keep them safe from predators, make sure the area is enclosed.

Activities for Enrichment

Serama hens need mental engagement to stay content and healthy. Include enriching activities in their surroundings. You can keep your hens interested with easy additions like safe toys, dust baths, and perches. For example, foraging behavior can be promoted by hanging goodies or vegetables.

Social Engagement

Serama hens are gregarious creatures that enjoy interacting with others. To encourage social behavior, keep them in small groups if at all possible. Because these birds are smaller than many other kinds, watch how they interact to make sure they are not being bullied.

Changes in the Environment

To avoid overgrazing and promote the growth of grass and plants, rotate outdoor spaces. Additionally, this method lessens the possibility of parasites building up in a small space. By altering their surroundings, you may support a natural, healthy ecosystem that will enhance their general well-being.

CHAPTER EIGHT

Taking Care Of Chickens In All Seasons

Growing Serama chickens can be a fulfilling experience, but it's important to know how to take care of them in the various seasons. The health and productivity of your flock may be impacted by the special opportunities and difficulties that come with seasonal changes. This article will offer crucial advice on how to manage summer heat, winterize the coop, and modify care for seasonal changes.

Adapting Care To Seasonal Variations

Your Serama hens' needs change with the seasons. Every season has its own set of difficulties, ranging from fluctuating daylight hours to extremely high or low temperatures. Maintaining the health and happiness of your flock requires an understanding of these variances.

Spring Cleaning

Spring is a season of growth and rebirth. Chickens become more active and produce more eggs when the temperature rises. Now is a wonderful time to inspect your coop and fix any issues that need to be fixed. As your hens forage more, make sure they have access to fresh water and think about supplementing their diet with grit to aid in digestion.

You can also experience an increase in pests like ticks and mites in the spring. Check your coop and hens frequently for indications of infestations. Think about applying natural pest repellents such as neem oil or diatomaceous earth. Additionally, adding a dust bathing space helps lower parasite burdens and helps your hens retain the health of their feathers.

Care for the Summer

Keeping your Serama chickens comfortable and cool during the summer will be your top priority.

Because of their vulnerability to heat stress, chickens may become lethargic and produce fewer eggs. It's important to provide shade, so think about adding awnings or trees to your chicken run. To keep your hens hydrated, make sure they always have access to fresh water.

You can also give them frozen treats made from fruits and vegetables or add ice cubes to their water on really hot days. These keep them interested in addition to assisting in lowering their body temperature. Keep a watchful eye out for symptoms of heat stress in your hens, such as panting or drooping wings, and intervene quickly if you see them.

Fall Maintenance

It's time to get your flock ready for the upcoming cooler months as summer draws to a close and autumn draws near. Chickens will want additional nourishment to accumulate fat stores for the winter, so keep feeding them high-quality feed. As the weather cools, it's also a good time to start cutting back on the amount of time your hens spend outside, particularly in the evening.

To keep your hens warm, check your coop for drafts and think about installing insulation. It's also crucial to give the coop a good cleaning before winter arrives. To avoid illness, sanitize the coop, remove old bedding, and clean the feeders and waterers. Lastly, to help your hens stay healthy as they get ready for winter, make sure they have access to grit and calcium supplements.

Winter Maintenance

For poultry keepers, winter can be the most difficult time of year. Winterizing your coop is essential if you want to give your flock a warm, secure space. Verify that there are no drafts and that the coop is well insulated. For this, insulating foam boards or straw bales can be used.

To help protect against the cold, supply a lot of bedding material, such as straw or shavings. For added warmth, you might also wish to install a heat lamp, but proceed with caution to avoid potential fire hazards. During this time of year, heated waterers can be a great purchase, so make sure your hens have access to unfrozen water.

The Coop's Winterisation

A crucial step in guaranteeing your flock's health during the winter is winterizing your Serama chicken coop. In addition to lowering stress and improving general health, proper planning can help maintain a secure and comfortable atmosphere.

The Coop's Insulation

Insulating your coop is the first step in winterizing it. Despite their resilience, chickens still need to be shielded from drafts and extremely cold temperatures. Examine the floor, roof, and walls for any openings that might allow cold air to enter. Cover exposed sections with stiff foam insulating boards or straw bales. Make sure the windows and doors close securely by checking them as well.

Control of Ventilation

Maintaining the coop's warmth is vital, but so is having enough ventilation to avoid moisture and ammonia accumulation, which can be detrimental to your hens' respiratory health. Make use of movable vents that can be opened or closed in response to changing weather. Make sure the ventilation is sufficient to prevent drafts from landing directly on the birds.

Areas for Nesting and Bedding

During the winter, it is crucial to have enough bedding in the coop. For comfort and insulation, use wood shavings or straws. Deep litter techniques, which involve letting bedding build up over time and producing heat as it breaks down, can be especially successful. Additionally, make sure nesting boxes are well-bedded to promote egg laying in the winter.

Food and Water Factors

Providing a high-quality feed is crucial because, in cold weather, chickens require additional energy to retain body heat. Additionally,

since they can produce heat while they digest, you might want to think about including warm treats like scratch grains or grains soaked in warm water.

Make fresh water that doesn't freeze accessible as well. You may keep water available and make sure your hens stay hydrated by using heated waterers or buckets with submersible heaters.

Keeping an eye on your flock

It's crucial to keep an eye out for any indications of stress or disease in your hens throughout the winter months. Observe how they behave and keep an eye out for any indications of frostbite, particularly on wattles and combs. To guarantee appropriate care, speak with a veterinarian if you observe any problems.

Controlling Summer Heat

To keep your Serama chickens healthy and productive throughout the summer, you must control the heat. These practical tips will help you keep your flock comfortable and cool.

Offering Shade

Making sure your hens have plenty of shade is one of the most crucial parts of managing summer heat. The best shade comes from trees, but if that isn't possible, think about covering their run with tarps or shade cloth. During the hottest times of the day, your hens will have a cooler place to hide out.

Staying Hydrated Is Essential

There should always be cool, fresh water available. Heat stress can cause chickens to rapidly become dehydrated in hot weather. Several water sources spread out around their neighborhood might promote drinking. Another option to keep dirt out of the water is to use a hanging waterer.

Consider filling the water containers with frozen water bottles or ice cubes to cool the water. This can help keep the temperature down and provide your hens with a cool beverage.

Enrichment and Cool Treats

To assist your chickens cope with the heat, give them cool snacks. A healthy and enjoyable method to keep children occupied and cool is with frozen fruits and veggies. Just cut up fruits, such as berries or watermelon, freeze them in ice cube trays, and then serve them as a cool snack.

Additionally, you can offer your hens with shallow pans of water to wade in. In addition to keeping them cool, this promotes their innate foraging and scratching habits. To avoid drowning, just make sure the water is shallow enough.

Modifying Feeding Procedures

Think about shifting your feeding schedule to the cooler hours of the day, such as early morning or late evening, on the hottest days. Your hens' metabolic heat output may rise if you feed them during periods of high heat. A balanced diet with lots of electrolytes should also be given, particularly if you observe symptoms of heat stress.

Observing Conduct

Watch your hens carefully for symptoms of heat stress, like as drooping wings, drowsiness, or panting. Take quick steps to calm them down if you observe any of these symptoms. Mild heat stress can frequently be reduced by providing fresh water and cool, shaded settings.

How to Get Ready for Heat Waves

Consider taking extra precautions if a heat wave is forecast, such as bringing your hens inside during the hottest hours of the day or using fans to move air around the coop. The comfort level of your chickens can be significantly increased by creating a breezy environment.

CHAPTER NINE

How To Care For And Train Serama Chickens

For your Serama hens to be socialized and well-adjusted, proper handling and training are crucial. These amiable little birds are well-known for their distinctive looks and endearing dispositions. When preparing your hens for contests or exhibitions, proper training and handling methods can strengthen your relationship with them.

Fundamental Handling Methods

Comprehending Serama Physiology

It's important to know your Serama hens' physical traits before managing them. Small and delicate, serama hens typically weigh less than a pound. Because of their lightweight frames, they are easily hurt if handled incorrectly. Always approach your birds with composure and gentleness, being sure to provide them with enough body support.

Getting Close to Your Chickens

To make your Serama hens feel comfortable, talk to them gently when you approach them. Steer clear of abrupt movements since they may frighten the birds. When picking up a chicken, cautiously reach out and let your hand become used to you. Using one hand to hold the weight of the bird from underneath, gently wrap it around its body. To avoid fluttering and possible harm, you might grasp the wings with your other hand.

Keeping Your Chickens in Check

After picking up your Serama, give its body complete support. To prevent tension, hold the chicken in your arms while maintaining a steady head and neck. The bird will feel more safe if you hold it close to your body. Gently put the bird back in its coop if it appears upset. To foster trust and increase your chickens' comfort level when being held, practice handling them regularly.

Getting Along With Others And Managing Your Chickens

Establishing a Convenient Setting

Making your Serama hens' surroundings pleasant is the first step in socialization. Make sure they have enough room to move about, perch, and interact with their environment. You should spend time with them every day so they can see you and learn your voice. Your hens will be less shy and more gregarious in a relaxed and friendly setting.

Techniques for Positive Reinforcement

You may effectively socialize and tame your Serama hens by using positive reinforcement. When they come up to you or react favorably to your voice, give them snacks like tiny pieces of fruit or vegetables. Over time, trust will grow as a result of them learning to identify your presence with positive experiences through consistent rewards.

Managing While Socialising

Increase the frequency and length of your handling sessions progressively as your hens get more accustomed to being handled. Talk to them in a soothing tone while gently stroking their feathers. Some chickens may take longer than others to warm up, so keep an eye on their responses and exercise patience. Patience and consistency are essential for effective socialization.

Getting Ready For Competitions Or Shows

Creating a Training Schedule

Establishing a regular regimen is essential to training your Serama hens for performances or contests. Allocate a specific period every day for training activities. Concentrate on particular abilities or habits you wish to cultivate, such as posing or leash walking. Their interest and attention will be maintained if the sessions are brief and interesting.

Promoting Desired Actions

During training, use positive reinforcement to promote desired behaviors. Give your chicken praise or goodies when it does a trick or acts in a way that pleases you.

Introduce increasingly difficult activities gradually, including leashed walking or staying motionless when instructed. Keep in mind that every bird learns differently, so be patient.

Getting Ready for Contests

Get your Serama hens used to the show setting before submitting them to competitions.

To introduce children to the sights and sounds they will experience, go to local performances. To make sure your chickens stay collected and quiet throughout the event, practice handling and posing them in a range of situations. On the day of the competition, your hens will feel safe and confident if you give them lots of support and encouragement.

CHAPTER TEN

Getting Ready And Competing

Getting Serama Chickens Ready For Performances

Overview of Show Preparation

For any novice farmer, preparing Serama chickens for shows is a fascinating adventure. Poultry fans are drawn to these little, attractive birds because of their distinctive appearance and endearing personalities. You need to pay attention to several things, such as training, grooming, and health, to make sure your Serama chickens are prepared for competition.

Nutrition and Health

Make sure your Serama chickens are healthy before submitting them to a competition. A balanced diet is essential to their general well-being. Give your hens premium commercial feed that has been specially prepared for them, as well as vitamins and minerals as supplements. To improve their diet, they can also incorporate fresh grains and veggies.

Frequent health examinations are essential. Keep an eye out for symptoms of disease, such as odd droppings, poor feather condition, or lethargy. Additionally, vaccination against common poultry diseases is crucial. To keep your flock healthy and prepared for the show, speak with a veterinarian who specializes in poultry.

Taking Care of Your Serama Chickens

One important part of getting ready for a show is grooming. Long before the competition, begin grooming your Serama chickens. A week before the performance, bathing them helps clean their feathers of oil and debris, bringing out their natural hues. Use a gentle shampoo made especially for birds, and make sure to rinse well to prevent skin irritation.

Use a towel or a low-speed blow dryer to gently pat dry your chickens after bathing. After drying, carefully comb their feathers to remove any tangles and make sure they are fluffy and attractive. Keep an eye on the legs and feet, cleaning them to avoid dirt accumulation that could detract from their appearance during judging.

Training and Socialisation

For your Serama hens to feel at ease in the show setting, socialization is essential. Frequent handling lessens their anxiety during competitions by acclimating them to being around people. To get them used to the environment they will be in, try gently arranging them on a show table.

It's also crucial to teach your Serama hens to pose and remain motionless. This conduct makes it easier for judges to assess their conformance. Reward your hens for remaining motionless by using treats to motivate them to stand in the desired position. They will gain composure and confidence with regular practice.

Comprehending Show Criteria And Evaluation

Overview of the Show Criteria

Success in competitions is largely dependent on knowing the show requirements for Serama chickens. While there may be particular regulations for each show, all judging is subject to a few broad guidelines. You may better prepare your birds for success by being familiar with these factors.

Physical attributes

Serama hens are judged according to their physical characteristics. Size, coloration, feather quality, and general body shape are important traits. Judges frequently search for birds that exhibit the compact stature that is characteristic of Serama chickens. To choose the greatest candidates for the exhibition, make sure you are aware of the recognized standards for various colors and patterns.

Behaviour and Temperament

Judges evaluate the birds' behavior and disposition in addition to their physical attributes. Serama chickens are renowned for being amiable, which has a big impact on their ratings. Throughout the tournament, make sure your hens are composed and behaving properly. A bird that exhibits signs of stress or aggression is unlikely to perform well when scrutinized by the judges.

The Process of Judging

Comprehending the judging procedure is essential for competition success. Usually, judges assess the birds methodically, beginning with their physical traits and progressing to their behavioral traits. Expect your chickens to be handled and inspected closely. They will perform better during judging if they are more used to being touched and inspected.

Advice On Winning At Contests
The Key Is Preparation

It takes careful planning to win at Serama chicken competitions. Make sure your birds fulfill the required criteria for grooming and health first. Judges will be more impressed and a well-prepared chicken will stand out from the competition.

Recognize the rules of the competition.

Rules and regulations vary from competition to competition. Learn these rules so that you won't be caught off guard on the day of the show. You can perform much better and handle the event more easily if you have this expertise.

Get there early and get settled.

To give your hens time to adjust to their new surroundings, arrive early on the day of the competition. Place them comfortably in their cages and give them food and fresh water. They will feel less stressed and more at ease if you let them unwind before the competition.

Learning and Networking

Talk to other showgoers who are interested in poultry. Networking can yield insightful information and advice from more seasoned rivals. This community can be a great resource for enhancing your chicken farming methods, so don't be afraid to share your experiences and ask questions.

Reflection After the Competition

Spend some time thinking back on the competition after it's over. Examine what went well and what needs to be improved. You will develop as a Serama chicken farmer and improve your show preparation with this introspective exercise.

CHAPTER ELEVEN

The Promotion And Sale Of Serama Chickens

Investigating New Markets For Your Chickens

Knowing your target market is essential when starting a chicken farm in Serama. Known for their small stature and distinctive look, Serama chickens are popular with collectors, enthusiasts, and pet-seeking families. You can more successfully customize your marketing efforts by identifying potential customers.

Collectors and Hobbyists

Poultry lovers and collectors who value the distinctive traits of the breed make up the main market for Serama chickens. These people are a niche market that is prepared to spend money on high-quality birds since they frequently attend poultry fairs and exhibits. Getting to know possible buyers through local and regional poultry exhibitions can be a great networking opportunity. You can raise awareness and interest in your breeding stock by showcasing your Serama chickens at these gatherings.

Pet owners and families

Families looking for unusual pets for their homes are another possible market. Serama chickens are a great option for families with kids because they are little, amiable, and rather simple to maintain.

You can capitalize on the growing trend of people keeping hens for company rather than just eggs by marketing your birds as family pets. These audiences can be successfully reached by distributing interesting content about the temperament and upkeep of Serama hens on social media and through neighborhood associations.

Internet-Based Marketplaces

Online marketplaces present a substantial opportunity for the sale of Serama hens in the current digital era. You can display your birds to a larger audience by using services like Craigslist, Facebook Marketplace, and specialized poultry markets. Make eye-catching listings that showcase your Serama hens' distinctive qualities, such as their temperament, size, and color differences. Adding thorough descriptions and excellent photos to your ads will improve them and draw in more potential customers.

Local Pet Expos and Farmers' Markets

Serama chickens can be sold directly to consumers at pet expos and farmers' markets. You can engage with prospective purchasers and respond to any inquiries they might have about the breed by setting up a stand at these events. Giving consumers information about housing, feeding, and care will reassure them and build your reputation as a breeder. To encourage clients to make an immediate purchase, think about providing promotional offers or bundle discounts.

Formulating Successful Advertising Plans

Reaching your target audience and promoting your Serama chicks requires effective advertising techniques. A comprehensive advertising strategy can be produced by combining traditional and digital marketing strategies.

Marketing on Social Media

Social media sites like Facebook, Instagram, and TikTok are excellent resources for promoting your Serama chicks. Make eye-catching content that highlights your birds' distinctive characteristics and includes videos of them interacting. By publishing frequently and answering messages and comments right away, you can interact with your audience. To network with possible buyers and exchange your knowledge, think about joining regional farming and poultry associations.

Development of Websites

Establishing a specific website for your Serama chicken farm can act as a focal point for sales and information. High-quality photos, thorough descriptions of your hens, maintenance instructions, and cost details should all be on your website. Customers can place orders straight online by including an e-commerce component. To increase website traffic, think about including a blog where you may post updates about your breeding program, advice on Serama care, and your experience.

Email Promotion

Email marketing is a powerful tool for keeping in touch with past clients and informing them of new products and discounts. Gather email addresses from customers at local events and on your website, and give them the option to sign up for updates.

Frequent emails can help you keep your audience interested in your birds by including updates about your farm, special deals, and care advice for Serama.

Local Promotion

Don't undervalue the influence of regional advertising. You can target local buyers by putting leaflets in veterinary clinics, community centers, and nearby feed stores. Working together with nearby companies can also lead to cross-promotional opportunities,

where you can promote each other's goods and services. Word-of-mouth advertising is still quite effective; encourages happy consumers to tell others about their experiences.

Regulations And Legal Aspects

Understanding the laws and rules about poultry production is crucial before selling Serama chickens. By being aware of these factors, you can make sure that your company runs morally and lawfully, safeguarding your interests as well as those of your clients.

Rules Regarding Zoning

It's important to confirm whether your property is zoned for chicken production because zoning laws differ depending on the area. The number of hens you can keep, the housing needs, and other limitations may be governed by local legislation. For comprehensive information to guarantee adherence to all rules, get in touch with your local government or zoning office.

Regulations for Health and Safety

Selling livestock, especially chickens, may be subject to certain health and safety laws in many places. Vaccinations, disease prevention strategies, and biosecurity procedures are a few examples. Learn the regulations in your area to make sure your hens are disease-free and healthy. Giving purchasers health certifications will increase your reputation and reassure them of the caliber of your food.

Licenses for Businesses and Sales Tax

Selling chickens could necessitate a business license and adherence to sales tax laws, depending on your location. To learn about the required permissions and tax duties for your operation, speak with a local accountant or business expert. Maintaining compliance with tax laws and the general financial stability of your company depends on keeping accurate records of your sales and spending.

Standards for Animal Welfare

Lastly, get knowledgeable about the animal welfare regulations that apply to the raising of chickens. In addition to meeting legal standards, making sure your Serama chicks are raised humanely and ethically will appeal to ethical customers. You may increase consumer trust and establish yourself as an ethical breeder by learning about the best standards for animal care.

CONCLUSION

For both novices and experienced farmers, beginning a serama chicken farming adventure can be one of the most fulfilling experiences. It's crucial to recognize the different aspects of serama chicken management, breeding, and care that you have learned when you consider the material covered in this book. The purpose of this conclusion is to emphasize the benefits of serama chicken farming, highlight the key lessons learned, and inspire you to learn more about this rewarding activity.

A Summary of the Main Ideas

We started by talking about the distinctive qualities that make serama hens different from other breeds. They are a pleasure to nurture and to see because of their small stature, endearing dispositions, and colorful feathers. You have greater respect for these small birds now that you are aware of their origins and the history of the serama breed.

Important subjects including housing, nutrition, and health management were examined in our investigation of serama chicken care. Since serama chickens do best in conditions that are clean, safe, and roomy, proper housing is essential. They have unique nutritional requirements, necessitating a well-balanced diet to promote their development and overall health. Preventive treatment and routine health examinations are essential to keeping your flock healthy and productive.

Accepting the Community

Keep in mind that you are a member of a wider community of poultry enthusiasts as you start your serama chicken farming adventure. Using social media, online forums, or local groups to interact with other serama farmers can yield insightful information and helpful assistance. In addition to increasing your knowledge, exchanging stories, advice, and best practices will help you make friends that will improve your farming experience.

Raising Serama Chickens Is Fun

Beyond their appearance, serama chickens are a source of joy to raise. Their amiable disposition and capacity to form bonds with their carers can infuse your everyday routine with coziness and warmth. It can be immensely satisfying to see them flourish, engage with them, and observe their amusing behaviors.

Serama hens are also frequently kept for their potential for exhibition. Participating in local competitions or exhibitions with your hens can give you a tremendous sense of accomplishment. These gatherings provide you the chance to demonstrate your diligence and commitment while also giving you the chance to pick up tips from locals.

Going Ahead

As your first learning phase comes to an end, think about establishing reasonable objectives for your serama chicken farming endeavors. A clear vision can help you make decisions going ahead, whether your goal is to start a modest breeding enterprise, compete in local shows, or just enjoy the company of these unusual birds.

Finally, keep in mind that raising serama chickens is a learning and development process. Accept the difficulties and acknowledge your accomplishments, no matter how modest. Serama chicken farming will bring you great delight if you have patience, perseverance, and a love for your feathered companions. I hope you had a rewarding experience and that your flock flourishes in your care!

THE END

www.ingramcontent.com/pod-product-compliance
Lightning Source LLC
Chambersburg PA
CBHW071409220526
45469CB00004B/1223